From Beet to Sugar

From Beet to Sugar

Ali Mitgutsch

 Carolrhoda Books, Inc., Minneapolis

LIBRARY OF CONGRESS CATALOGING IN PUBLICATION DATA

Mitgutsch, Ali.
  From beet to sugar.

    (A Carolrhoda start to finish book)
    Ed. for 1972 published under title: Von der Rübe zum
Zucker.
    SUMMARY: Highlights the step-by-step process of
extracting sugar from sugar beets.

    1. Beet sugar—Juvenile literature. [1. Sugar] I. Title.

TP390.M5713   1981   664′.123   80-29603
ISBN 0-87614-145-9

        2 3 4 5 6 7 8 9 10 86 85 84 83 82

From Beet to Sugar

Sugar is made from two kinds of plants:
sugar cane and sugar beets.
This is the story of how sugar is made
from sugar beets.
Sugar beets grow beneath the ground.
When they have finished growing,
the farmer uses a machine
to cut off the tops of the plants.
Then he plows up the sugar beets.

When he has plowed up all of his beets,
he puts them into his truck
and takes them to a sugar factory.

At the factory the sugar beets are washed.
Then they are put into machines
which cut them into small pieces.
The small pieces are called **chips**.

The chips are then put into containers
full of hot water.
The hot water will soak the sugar out of the chips.
There is a drain at the bottom of the container.
The sugar syrup will pass through this drain.
The chips will be saved and used for animal feed.

Next the sugar syrup is put through a filter to clean it.

Then it goes into a large pot where it is boiled.

Most of the water boils away.

Now the sugar looks like thick brown mush.

This mush is called **molasses** (muh-LASS-uhz).

The molasses is put into a machine
called a **centrifuge** (SEHN-trih-fewj).
The centrifuge spins around very fast.
It separates the liquid that is
still in the molasses from the sugar.
Now the sugar looks like the sugar we buy.
It is made of many tiny white crystals.

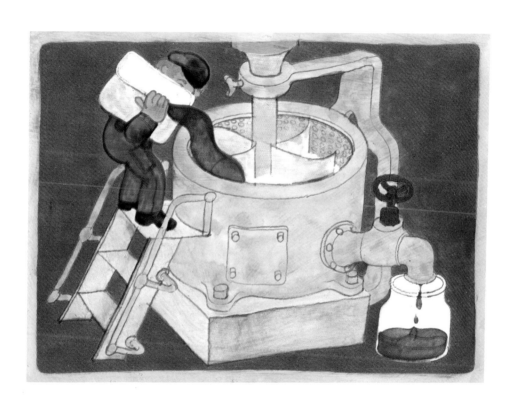

At last the sugar is ready to be put into bags.

Some of it will be pressed into sugar cubes.

Then the sugar will be sent to stores.

Many foods that taste sweet have sugar in them.
Too much sugar can give people cavities
and make them fat.
But a little sugar makes many foods taste delicious.

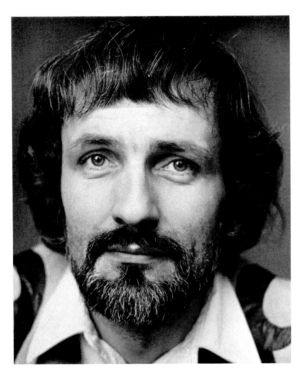

Ali
Mitgutsch

ALI MITGUTSCH is one of Germany's best-known children's book illustrators. He is a devoted world traveler, and many of his book ideas have taken shape during his travels. Perhaps this is why they have such international appeal. Mr. Mitgutsch's books have been published in 22 countries and are enjoyed by thousands of readers around the world.

Ali Mitgutsch lives with his wife and three children in Schwabing, the artists' quarter in Munich. The Mitgutsch family also enjoys spending time on their farm in the Bavarian countryside.

THE CAROLRHODA

START

TO FINISH

BOOKS